# Table of Contents

# 1. Introduction

## 1.1 About the Author

*Hey there, I'm a Chemical Process Engineer with a passion for learning, teaching and sharing my knowledge. I got my degree in Chemical Engineering from the Technical University of Denmark and the University of British Columbia, Vancouver.*

*The purpose of this handbook is to provide you, as a chemical process engineering student or young professional, with some of the most foundational competencies and problem-solving methods frequently required in the first 3-5 years of your real-life engineering career.*

*During my studies I was frustrated with a lot of the heavy engineering books we had to read, because of the amount of unnecessary information packed into each chapter and the obsession with academic wording and sentence structuring. I remember thinking to myself back in the day: There must be a simpler way of conveying these concepts and methods!*

*My goal is to create valuable down-to-earth study material & handbooks for applied engineering. For this book, I have chosen the topic of applied mass balances and systems based on my real-world engineering experience as it proved to be the most valuable skill in my toolbox coming out of university. The understanding of systems and balances presented in this book, also provided me with a stepping stone to deeper learning in the subjects of fluid- and heat-transfer.*

*Thanks for purchasing this book on Chemical Process Engineering. I'm excited to share my knowledge and experience with you and hope it will be helpful for your learning and professional development.*

*Humbly and gratefully,*

Michael Kay Hoffmann

# 1.2 Overview of the Content

This guidebook is for process engineers looking to master the principles of mass balances. The book is divided into three core chapters, each building on the previous one, to provide the reader with a thorough understanding of the subject.

Chapter 2 introduces the common process variables, concepts, and symbols used in mass balance calculations. It also provides an introduction to flow diagrams and covers systems and states.

Chapter 3 delves into mass balances for simple and steady-state systems. It covers the principles of balances, steady-state systems, and takes the reader through practical examples of solving these systems. This chapter also provides a systematic step-by-step approach to solving mass balances.

Chapter 4 explores mass balances for complex and non-steady-state systems. It covers complex systems with recirculation and bypass, and non-steady-state systems with time dependent variables. This chapter also includes practice examples that equips the reader with the skills needed to perform mass balances on real-life installations.

Overall, this book is an essential resource for process engineers looking to learn or enhance their skills in mass balance calculations. The practical examples and assignments provided throughout the book makes it a useful tool for self-study and reference in the role as a process engineer.

# 1.3 Process Engineers: Where Do They Work? What Do They Do? What Skills Do They Need?

The chemical process industries are a diverse set of industries that involve the processing and transformation of raw materials into value-added products. Process engineers play a critical role in these industries by designing, optimizing, and managing the various unit operations and processes that make up these industries.

The chemical process industries encompasses a wide range of industries, including:

- Production of basic chemicals
- Specialty chemicals manufacturing
- Polymer manufacturing
- Petrochemical manufacturing
- Pharmaceutical manufacturing
- Food and beverage production
- Consumer goods manufacturing
- Energy production

Process engineers may be involved in a variety of tasks, such as designing and operating chemical- and biochemical reactors, distillation columns, and other process equipment. They may also be responsible for developing and implementing process control strategies, conducting laboratory experiments to evaluate and optimize processes on cost and yield, and troubleshooting process issues. They are also typically involved in the health, safety and environmental activities because of their deep understanding of the production and manufacturing processes.

Process engineers must be able to balance technical knowledge with business acumen, as they are often responsible for ensuring that their processes are both time-efficient and cost-effective.

In order to succeed as a process engineer, it is essential to have a strong foundation in key concepts such as mass and energy balances, stoichiometry, chemical equilibrium, and chemical kinetics. These topics are fundamental to many of the day-to-day tasks performed by process engineers, including designing and optimizing chemical processes, troubleshooting problems, and ensuring safety and regulatory compliance.

Understanding systems and being able to apply mass balances are the very foundations of process engineering and are critical stepping stones to applying energy balances, fluid-mechanics, biology, physics, and chemistry as a professional process engineer.

Process engineers are usually extremely valuable for the companies that employ them. Their technical background and focus on understanding complex processes make them a good all-around asset for a wide range of tasks and challenges. When something goes wrong in production, process engineers are a must-have for a detailed root-cause analysis, and when things are running smoothly, process engineers are needed to keep them that way.

Process engineers have a variety of roles and responsibilities in different industries. Here are some of the many roles that a process engineer can undertake:

**Process Design**: Process engineers design and develop new processes or modify existing ones to improve efficiency, product quality, and safety. They work with project teams to identify technical requirements, develop process flow diagrams, and design process equipment.

**Process Control:** Process engineers develop control strategies to optimize the performance of processes, such as temperature, pressure, flow, and composition. They work with automation specialists to implement control systems, and troubleshoot issues to improve production efficiency and reduce costs.

**Process Improvement**: Process engineers analyze data to identify areas for process improvement, develop process improvement plans, and implement process changes. They work with cross-functional teams to identify and evaluate improvement opportunities, and develop solutions to meet business needs.

**Research and Development**: Process engineers work in research and development to develop new products and processes, and improve existing ones. They design experiments, analyze data, and provide recommendations to improve process efficiency and quality.

**Process Safety**: Process engineers are responsible for ensuring the safety of the process, equipment, and personnel. They conduct risk assessments, develop and implement safety procedures, and train personnel on safe operating practices.

**Environmental Compliance**: Process engineers ensure that the process and equipment meet environmental regulations, such as air emissions, waste disposal, and water treatment. They work with environmental specialists to develop compliance strategies and implement solutions to mitigate environmental impact.

**Project Management**: Process engineers manage projects from concept to completion. They develop project plans, track project progress, manage budgets, and ensure that project deliverables meet technical and business requirements.

Given this long and in-complete list of potential roles and problems that a well-rounded process engineer can tackle, it is obvious that a process engineer must have a diverse set of skills and knowledge in order to be successful.

They must be able to understand the fundamental principles of chemistry, biology, physics, and mathematics, as well as be proficient in the use of process simulation software and other engineering tools. In addition, a process engineer must have strong communication skills in order to effectively collaborate with other engineers, project managers, and clients.

They need to be able to think critically and creatively to come up with innovative solutions to complex problems.

Overall, a successful process engineer is a versatile problem-solver with a vast set of skills that are all rooted in understanding systems and balances within those systems.

# 2. Symbols, Systems & States

Chapter 2 will introduce the most common process variables, associated units, and block flow diagrams. After this chapter, the reader should have achieved the following competencies:

- ❖ Calculating mixture composition expressed in mole or mass fractions and converting between the two.
- ❖ Calculating the volume of a gas by utilizing the ideal gas law.
- ❖ Drawing a block flow diagram (BFD) for a production process based on the process description and setting up system boundaries for use in setting up mass balances.
- ❖ Understanding and identifying the different types of systems and states of systems.

## 2.1 Common Process Variables, Concepts and Symbols

In this section several essential process engineering concepts and variables that are referenced throughout the book will be defined:

| Terms of Reference | | Formula/Ex. | Units |
|---|---|---|---|
| Component | A pure chemical compound in a mixture or stream | $CH_4$, $O_2$, $CaCO_3$, $H_2O$, NaCl, $SiO_2$ | - |
| Element | A chemical substance that cannot be broken down into other substances | All elements of the Periodic Table e.g.: Cl, He, Na, O, C | - |
| Mass | Weight of a given component i | $m_i$ | kg |
| Mass flow | Flow of mass of a given component i | $m_i$ (denoted in weight/time) | kg/s |
| Total mass/mass flow | The combined weight or weight flow of all components of a stream or system | $\sum m_i$ | kg or kg/s |
| Mass fraction | The mass of a component i, divided by the total mass of a mixture/stream | $w_i = \dfrac{m_i}{\sum m_i}$ | dimensionless |
| Weight% | Mass fraction multiplied by 100, to express the fraction on percent-basis | $w\%_t = w_i \times 100$ | dimensionless , % |

| Terms of Reference | | Formula/Ex. | Units |
|---|---|---|---|
| Mole | The amount of substance of a component i, that contains as many elementary parts as the number of atoms in 12 grams of carbon-12 | 1 mole of i | mol |
| Amount/flow rate of substance | The number of moles of a mixture or stream of component i | $n_i$ | mol or mol/s |
| Molar mass | weight of 1 mole of component i | $M_i$ | kg/mol |
| Mole fraction | The number of moles of component i, divided by the total moles of the mixture/stream | $y_i = \dfrac{n_i}{\Sigma n_i}$ (gas) $x_i = \dfrac{n_i}{\Sigma n_i}$ (liquid) | dimensionless |
| Mol% | Mole fraction multiplied by 100, to express the fraction on percent-basis | $x_i, y_i \times 100$ | dimensionless, % |
| Volume | The space occupied by an object. | V | $m^3$ |
| Volumetric flow | Flow rate measured in volume per unit of time. | $\dot{V}$ | $m^3/s$ |

| Terms of Reference | | Formula/Ex. | Units |
|---|---|---|---|
| Vol% | The volume of a component i, divided by the total volume of a mixture or stream multiplied with 100. | $\dfrac{V_i}{V_{total}} \times 100$ | dimensionless, % |
| Concentration | Amount of component i present in a given total volume or mass mixture or stream | Ci | mol/m$^3$ mol/kg kg i/kg mixture |
| Density | Mass of a component per volume of the same component | $\rho$ | kg/m$^3$ |
| Pressure | The amount of force per unit surface area | P | Pa (N/m$^2$) |
| Temperature | A measure of kinetic energy of particles in an object. | T | K |
| Ideal gas constant | A physical constant that relates P, V, T to the number of to the number of moles of a substance. | R | $J/(mol \times K)$ |
| Normal volume (Normal Cubes) | Volume of a gas at standard temperature and pressure (STP: 273.15 K and 101.3 kPa[1]) | Nm3 | m$^3$ |

[1] Other definitions are also used in the industry. For example: IUPAC [10] STP uses 1 bar instead of 1 atm.

In addition to this table let's briefly introduce the **ideal gas law** and how to use it to find the normal volume $Nm^3$:

An **ideal gas** is a gas, where intermolecular forces can be disregarded. This assumption creates a rather favorable equation for making very useful calculations on the behavior of the gas and its associated parameters. The ideal gas law is expressed as:

$$PV = nRT$$

No gasses are actually ideal, but all gasses get closer to ideal gas behavior/state at lower pressures and higher temperatures.

For ideal gasses two important relationships holds true:

$$volumetric\ fraction = mole\ fraction$$
$$vol\% = mol\%$$

This relationship is true since V and n are proportional, since P and T are equal for all components in a gas mixture/stream.

**Normal volume (*or Normal Cubes - Nm³*)** is a common way of describing an amount of gas in the process engineering industry. Using the Ideal Gas Law, if the amount of moles of gas is known, the amount of $Nm^3$ can easily be calculated if needed to report in this format to internal or external stakeholders or for comparison with manufacturing documents.

For 1000 mol of an ideal gas the equation looks like this:

$$V = \frac{nRT}{P}$$

$$V = \frac{1000 \ mol \times 8.315 \ \frac{m^3 Pa}{mol \ K} \times 273.15 \ K}{1.013 \times 10^5 \ Pa} = 22.42 \ Nm^3$$

With these key concepts and symbols in mind, let's move on to flow diagrams to establish how an engineer can visually represent a system.

# 2.2 Introduction to Flowsheets / Flow Diagrams

Flowsheets are essential tools used by process engineers to depict the flow of materials and energy throughout a process. A flowsheet provides a visual representation of a process and is often used as a starting point for the development of material and energy balances.

This subchapter will introduce the concept of flowsheets, the symbols used in their creation, and their importance in the chemical process industries. Additionally, we will cover the different types of flowsheets and the information that can be gleaned from their use. Finally, we will provide examples of flowsheets for various processes and discuss how to interpret them.

There are several types of flowsheets, including:

**Block Flow Diagrams (BFDs)**: BFDs provide an overview of the process, showing major equipment, process streams, and process conditions.

**Process Flow Diagrams (PFDs)**: PFDs are more detailed than BFDs, and include information on equipment, process streams, process conditions, and process control.

**Piping and Instrumentation Diagrams (P&IDs)**: P&IDs show the piping and instrumentation details of the process. These diagrams include information on valves, pumps, instruments, and control systems.

PFDs and P&IDs are outside of the scope of this book.

**Block Flow Diagrams (BFDs)** are the simplest type of diagrams that a Process Engineer will draft. They are usually used in the early stages of the design, optimization, or calculation phases of a project due to their simple yet effective way of designing system boundaries for systems and streams. In a block diagram, an arrow represents a stream, and a block represents a subsystem, unit or set of units where some form of treatment takes place.

Below is an example of a combustion process:

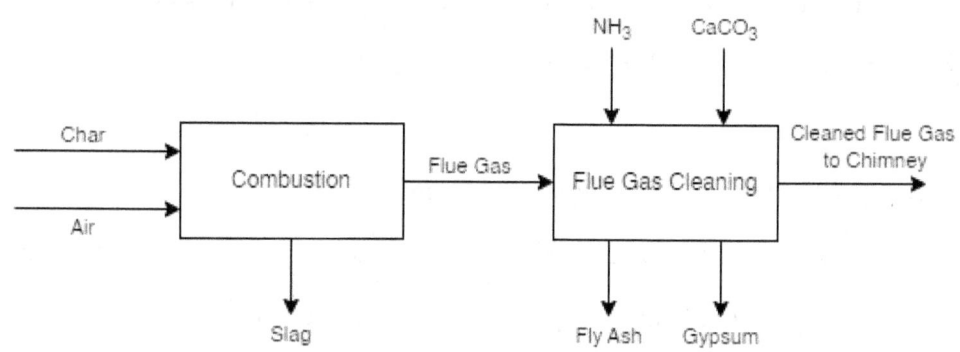

Figure 1: Block flow diagram of a char combustion process.

Two important take-aways for construction of best-in-practice BFDs:

1. **One stream = One arrow.** A stream can consist of multiple components, but never combine two streams into one arrow for sake of convenience. If there needs to be two streams in the physical world, this detail needs to be present already at the BFD-level of design. BFDs lay the groundwork for more detailed PFDs and P&IDs.

2. For simplicity and speed **keep the blocks simple**. As soon as we are using more detailed representation of the subsystems and process-units, the drawing will fall into the PFD-category. Blocks are simple and quick to draw, and it conveys the overview needed at this stage of the design and/or calculations.

Quickly drafting simple BFDs is often used to set up mass balances for yield calculations or to calculate size and composition of unknown streams. For this we need to add some information to our streams, such as known and unknown compositions and flow. The BFD below is an example of needed factors added to the BFD in order to assist the balancing and calculation process:

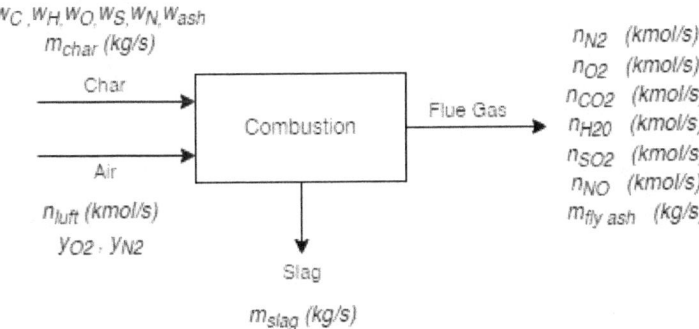

Figure 2: BFD of char combustion process with streams and composition information added.

In order to solve for $n$ unknown stream sizes and/or compositions we need to utilize mass balances to produce n independent equations. The step-by-step approach to do this is detailed in subchapter 3.6.

## 2.3 Systems & States

For any calculation on streams, it's vital to keep account of the total mass as well as the mass of the different elements and components. These accounts are called *balances* and we set them up for different *systems*. **Systems** are the part of the world we are interested in. For different objectives it can be a single unit in a factory, a multitude of units or the whole facility/plant.

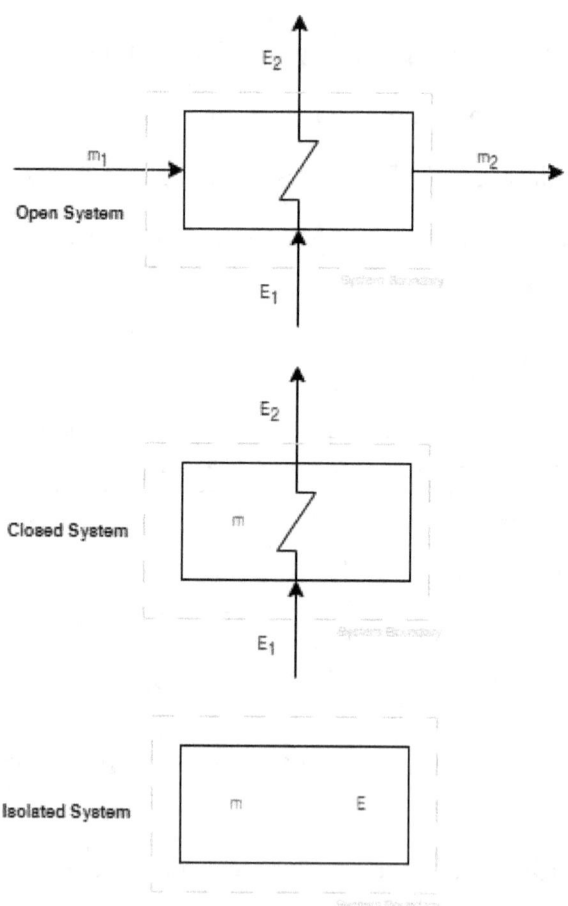

Figure 3: Open, closed and isolated systems. The zigzag line resembles energy transfer between the system and the environment.

**System boundaries** are defined to outline and limit the part of the world we are interested in (the system) from the rest of the world (environment). In process engineering we encounter three different kinds of systems as seen in figure 3.

These three types of systems can be in either steady state or non-steady state. Identifying the state of the system is key in how the mass balances are set up and solved.

On the following page is a table defining the three different systems and two different states:

| Systems | |
|---|---|
| Open | A system where mass is transported across its system boundary. |
| Closed | A system where no mass is transported across its boundary. Energy can still be transported in and out of this type of system. |
| Isolated | A system where no mass or energy is transported across its boundary. |
| **States** | |
| Steady | If a system is constant over time, it is in steady state. This means that the amount of mass and energy flowing into the system is equal to that of which is leaving the system. There is no accumulation of matter and/or energy in the system over time. |
| Non-steady | Counter to the steady state system - the system is changing over time. There is accumulation or loss of matter and/or energy over time. |

These concepts of types of systems and state of systems are important for all process engineers to grasp and will be used throughout this book. With these concepts in mind, let's explore the **accounting** of process engineering: **Balances.**

# 3. Mass Balances for Simple & Steady State Systems

This chapter will teach the principles for setting up and solving mass balances for total mass, elements, components in different systems; these concepts will be introduced and applied through practical examples and assignments. After this chapter the reader should have achieved the following competencies:

- ❖ Being able to describe the principles of mass-balances.
- ❖ Setting up and solving mass balances for steady state systems with and without reactions.
- ❖ Defining and utilizing the concept of key elements in setting up and solving mass balances.
- ❖ Understanding key principles of combustion processes and gas compositions.

## 3.1 Principles of Balances

To make important calculations and designs for any system the total mass, mass of a given element or component for a given system within a timeframe needs to be accounted for. For continuous processes the flow of energy and matter is usually tracked per second or hour. For batch processes the time frame is usually from start to end of the process itself.

Balances are built from the same general equation:

$$IN + PRODUCED = OUT + ACCUMULATED$$

When working with mass balances for the total system or component/element basis, these terms above will either be in units of mass or units of mass per unit of time (using the metric system, usually kg and kg/h).

When doing component- or element balances, where all terms will represent the same component or element the equation can be optionally set up with either mass or amount of substance (moles) since the molar mass (conversion factor between the two units) is the same for every term.

For total mass balances, where each term is a mix of all several components, the terms are all in units of mass. For the term *produced* the following applies:

- ❖ With total mass- and element balances the *produced* term will be equal to 0, as mass and individual elements cannot be created or consumed
- ❖ Components or chemical compounds can be converted through chemical reactions.
  - ➤ For components that are created (products of the reactions) the term *produced* will be positive ($> 0$).
  - ➤ For components that are consumed (reactants of the reactions) the term *produced* will be negative ($< 0$)

# 3.2 Steady State Systems without Chemical Reactions

For processes in steady state systems with no chemical reaction the mass balance is simplified as follows:

$$IN \ + \ PRODUCED \ = \ OUT \ + \ ACCUMULATED$$

Steady state entails that the term *accumulated* is equal to 0.

No chemical reaction entails that the term *produced* is equal to 0.

In this simple example the mass balance becomes very aesthetically pleasing:

$$IN \ = \ OUT$$

Let's kick off the practical part of this book with a simple example where two streams of sulfuric acid with different concentrations are mixed:

# 3.3 Practical Example: Mixing of Sulfuric Acid

**Problem statement:**

In a continuous process two streams of sulfuric acid with different concentrations are mixed in a mixing tank.

Stream 1:          199 kg/h, 23 w% $H_2SO_4$.

Stream 2:          301 kg/h, 96 w% $H_2SO_4$

Assumption:     Process is in steady state.

Calculate the mass flow and concentration of the leaving the mixing tank.

**Solution:**

When solving tasks related to mass balances it's always a good idea to sketch out the process in a simple BFD (Block Flow Diagram) and designate known and unknown variables.

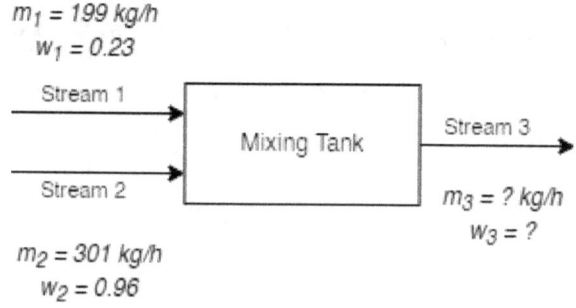

Figure 4: BFD - Mixing of sulfuric acid.

Beginning to see the power of simple BFDs? They are fast to draw and give a great overview of the system of interest.

Since there are no chemical reactions and the process is in steady state, we use the simplified version of the total mass balance:

$$IN = OUT$$

$$m_1 + m_2 = m_3$$

$$199 \, kg/h + 301 \, kg/h = 500 \, kg/h$$

*For this example this seems overly complicated for a simple calculation. However, this thought process is necessary when doing calculations on bigger and more complex systems.* Now that we know the mass flow of stream 3, we can find the composition by leveraging the component balance for $H_2SO_4$. The system is in steady state and there are no chemical reactions in this simple mixing example.

$$PRODUCED, \, ACCUMULATED = 0$$

$$IN = OUT$$

$$m_1 \times w_1 + m_2 \times w_2 = m_3 \times w_3$$

Substituting in the information we were given and the total mass flow:

$$199 \, kg/h \times 0.23 + 301 \, kg/h \times 0.96 = 500 \, kg/h \times w_3$$

Using a solver or some simple algebraic rearranging:

$$w_3 = \frac{(45{,}77 \, kg/h + 288{,}96 \, kg/h)}{500 \, kg/h} = 0.67$$

This means that 67 % of stream 3 consists of $H_2SO_4$.

It's always good to add the units to the calculations made to check if the end-result has the correct units. In this case the units of mass flow (kg/h) are divided out with each other and we are left with a unitless 0.67, which is what we were looking for: a mass fraction.

---

Summary of the steps taken to get the answer:

1. Set up the total mass balance to find the unknown mass flow
2. Set up the component balance for $H_2SO_4$ to find the composition of the unknown stream.

Let's do one more example on simple, steady state systems with no chemical reactions, where the two unknowns are linked to two separate streams.

# 3.4 Practical Example: Mixing of Sulfuric Acid (II)

Let's take the same example as before, but now with both inlet flow rates being unknown.

**Problem statement:**

Stream 1:      unknown flow rate, 23 w% $H_2SO_4$

Stream 2:      unknown flow rate, 96 w% $H_2SO_4$

Stream 3:      500 kg/h, 67 w% $H_2SO_4$

Calculate the mass flows of the two entering streams.

**Solution:**

Again, start by sketching a BFD and add the known and unknown variables:

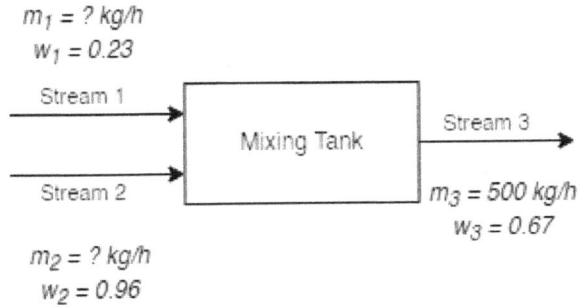

Figure 5: BFD - Mixing of sulfuric acid (II)

In this example we cannot immediately calculate the answer in our head. The reason for this is that our unknown variables are linked to two different streams. The problem is different, but similar and the recipe is principally the same. Since

26

we have two unknowns, we need to set up two linearly independent algebraic equations. Let's setup the total mass balance:

$$IN = OUT$$

$$m_1 + m_2 = m_3$$

and the component balance for $H_2SO_4$:

$$IN = OUT$$

$$m_1 \times w_1 + m_2 \times w_2 = m_3 \times w_3$$

Substituting in the known variables:

$$m_1 + m_2 = 500\,kg/h$$

$$m_1 \times 0.23 + m_2 \times 0.96 = 500\,kg/h \times 0.67$$

The result is two linearly independent equations and we solve these with direct substitution:

$m_1$ is isolated in the total mass balance and inserted into the component balance:

$$m_1 = 500\,kg/h - m_2$$

$$(500 \, kg/h - m_2) \times 0.23 + m_2 \times 0.96 = 500 \, kg/h \times 0.67$$

Now we can solve for $m_2$:

$$m_2 = 301 \, kg/h$$

$$m_1 = 500 \, kg/h - 301 \, kg/h = 199 \, kg/h$$

---

In cases of needing to solve more than two linear equations and unknowns it can be favorable to leverage matrices. Numerical solutions to non-linear equations are usually better solved using computer programs like Matlab. These calculations can however be simplified by setting up the right balances in a logical sequence. This systematic approach will be covered in the following section.

# 3.5 Setting Up & Solving Mass Balances

The method we used to solve mass balances in previous examples can be generalized and applied to more complex systems. One concept to grasp that can be massively helpful in solving these problems is the concept of Key Elements.

**Key Elements** are components that do not undergo a chemical reaction in a given system and which only exist in one ingoing and outgoing stream. A classic example of this in many process engineering systems are inert gasses such as Ar, He, and $N_2$ (under certain conditions). Another key element that is also considered in solving combustion processes is the carbon element. Usually, this will only exist in the fuel stream and the flue gas stream. In many processes, individual elements function as key elements. Problems that can be solved by setting up balances for key elements are also solvable by the general method, but the solution can be massively simplified if the balances are first set up for key elements.

Let's dive into a more systematic and generalized approach to solve these problems.

# 3.6 The Systematic Approach to Solving Mass Balances

**Step 1:** Define the problem. What information is known, and what do we seek to know? Is a problem that can be solved by setting up mass balances?

**Step 2:** Visualize the problem by sketching a block diagram and adding all known and unknown (if sought) variables, defined with either their numerical value or their symbols.

**Step 3:** Add SI units to every symbol and value. Make sure the units are consistent i.e. have all flow rates on the same weight and time basis (kg/h, g/s, tonnes/day etc.), so that they can be compared and combined properly. Convert mole flows into mass flows if needed using the molar mass of the component/element.

**Step 4:** When setting up the balances, the chosen system needs to be clearly outlined. Is it one unit or several units, a tie-in point in the piping or the entire facility? This should be visible from the simple BFD drawn in Step 2.

**Step 5:** If the system under consideration involves chemical equations, make sure to write up the equations as these are needed to calculate the *Produced* term of the balances where reactants and products of the reactions are involved.

**Step 6:** The required system of equations to solve a given system is achieved by two types of equations:

Mass balances on total mass, components and/or elements.

Condition-based equations for each stream:

$$\sum y_i = 1, \sum x_i = 1, \sum w_i = 1$$

Other condition-based equations can be given from the problem description.

**Step 7:** Start by setting up balances for key elements. A key element is a component that is not converted chemically in the system, which exists in only one incoming and one outgoing stream. A key element can also be an element that also exists in only one ingoing and one outgoing stream. These balances can often (but not always!) be solved independently of the other balance equations.

**Step 8:** During the calculations keep careful control of the units.

**Step 9:** Finish the calculations by checking that all balances are still consistent. One way to check if the balances are correct is to set up a control balance of another component that was not needed to solve the problem.

This recipe can be used as a guide to optimize for time and reduce the amount of potential mistakes that can be made when solving mass balances.

# 3.7 Mass Balances of Combustion Processes

In the following section mass balances for combustion processes will be explored. As combustion is a chemical reaction, this is a good introduction to solving systems that include chemical reactions.

Combustion processes are especially interesting because of its widespread usage in engineering due to its speed and vast amounts of energy released in the process. Most of the energy consumed today is still generated from combustion processes. For this chapter, let's dive into the concepts and mass balances of combustion processes.The most utilized fuels are char, oil, natural gas, bio-fuels, trash and LPG.

| Fuels | |
|---|---|
| Char | Mainly consists of carbon, hydrogen, oxygen, sulfur and nitrogen along with several non-combustible parts which turn to ash in the process. |
| Oil | Mainly consists of heavy hydrocarbons (fuel oil) or lighter hydrocarbons (gas oil or diesel) along with a minor sulfur component. |
| Biomass | An umbrella term and covers fuels such as wood, straw, kernels, rest products from agriculture such as manure and more. |

| Fuels | |
|---|---|
| Trash/Waste | Household- or industrially sourced waste is usually combusted to produce district heating and electricity. |
| Natural gas | Mainly consists of methane and minor parts ethane, propane and butane. |
| LPG | Liquefied Petroleum Gas consists of propane and/or butane, which is stored in liquid form under high-pressure to conserve space. This is often used as starter-gas for boilers or other combustion units to get a pilot flame going. |
| Types of combustion | |
| Complete | Complete carbon content of the fuel is converted to $CO_2$ during the combustion process due to sufficient oxygen being present. |
| In-complete | CO (carbon monoxide) is produced instead of $CO_2$ due to the limited amount of available oxygen as reactant in the combustion process. |

For combustion processes in the industry, the necessary oxygen is usually provided via ambient air. Dry air has the following composition:

| Dry Air Composition | |
| --- | --- |
| Component | Volumetric % |
| N2 | 78.08 vol% |
| O2 | 20.95 vol% |
| Ar | 0.93 vol% |
| CO2 | 0.03 vol% |

and trace amounts of Ne, He, Kr, Xe, $H_2$, $CH_4$ and $N_2O$.

When it comes to combustion processes $N_2$ and Ar are considered inert gasses. $CO_2$ is an insignificant part of ambient air, so to simplify calculations performed in this book the following composition of ambient air is considered accurate enough for this book:

| Dry Air Comp. (simplified) | |
| --- | --- |
| Component | Volumetric % |
| N2 | 79 vol% |
| O2 | 21 vol% |

**Nitric Oxide (NO)** can be produced from a part of the fuel's nitrogen content, and at temperatures over 1500°C the $N_2$ and $O_2$ from ambient air can react and form NO. This formation can be significant from an emissions-standpoint, but is insignificant when looking at mass balances and will not be considered in this book.

**Ambient airs molar mass** then comes out to:

$$M_{air} = y_{N_2} \times M_{N_2} + y_{O_2} \times M_{O_2}$$

$$0.79 \times 28\ g/mol + 0.21 \times 32\ g/mol = 28.84\ g/mol$$

$$M_{air} = 29\ g/mol$$

Most fuel consists of hydrogen, and therefore the resulting flue gas from combustion will include $H_2O$. Therefore the composition of a flue gas is often referred to as being on wet- and dry basis respectively:

**Wet basis** of a flue gas composition means that the flue gas holds water vapor.

**Dry basis** of a flue gas is often good to calculate, since water vapor is typically condensed from the flue gas before its measured, since this usually happens at normal ambient temperatures and pressures. It is therefore important to be able to convert between wet- and dry basis when looking at flue gas composition.

Let's take a look at how to convert between these two types of compositions.

# 3.8 Practical Example: Converting Composition from Wet- to Dry Basis

**Problem statement:**

A given flue gas on wet basis consists of:

65 mol% $N_2$, 15 mol% $CO_2$, 15 mol% $H_2O$ and 5 mol % $O_2$

Find the sum of the mol% without water.

**Solution:**

$$\Sigma y_{non-water} = 0.65 + 0.15 + 0.05 = 0.85$$

Composition on dry basis is then found by normalizing with the dry total mole:

$$y_{N_2, dry} = 0.65/0.85 = 0.765$$

$$y_{CO_2, dry} = 0.15/0.85 = 0.176$$

$$y_{O_2, dry} = 0.05/0.85 = 0.059$$

$$\Sigma y_{dry} = 0.765 + 0.176 + 0.059 = 1$$

# 3.9 Practical Example: Converting Composition from Dry- to Wet Basis

**Problem statement:**

A flue gas analysis gives the following composition on dry basis:

75 mol% $N_2$, 20 mol% $CO_2$ and 5 mol% $O_2$. A direct moisture reading of the water content in the flue gas gives a mole fraction for $H_2O$ of 0.11.

**Solution:**

$$\sum y_{dry} = 1 - y_{H_2O} = 1 - 0.11 = 0.89$$

$$y_{N_2,wet} = 0.75 * 0.89 = 0.67$$

$$y_{CO_2,wet} = 0.20 * 0.89 = 0.18$$

$$y_{O_2,wet} = 0.05 * 0.89 = 0.04$$

$$y_{H_2O,wet} = 0.11$$

$$\sum y_{wet} = 0.67 + 0.18 + 0.04 + 0.11 = 1$$

Let's introduce a few more concepts before diving into mass balances on systems with chemical reactions.

**Theoretical oxygen** is the number of mole $O_2$ (batch process) or mole flow of $O_2$ (continuous process) required for complete combustion of the fuel components C, H, S into $CO_2$, $H_2O$ and $SO_2$. Note that if the fuel consists of oxygen, the theoretical oxygen required is reduced proportionally.

**Theoretical air** $L_{theory}$ or $L_{min}$ is the amount of air that holds the theoretical oxygen.

**Practical air** L is the actual used amount of air in the process.

**Excess air coefficient** $\lambda$ is the ratio of amount of practical air to the theoretical air:

$$\lambda = L/L_{min}$$

These concepts will be used in the upcoming chapter and calculations of balances of steady state systems with chemical reactions.

# 3.10 Steady State Systems with Chemical Reactions

In systems where chemical reactions occur, the reaction equations are used to set up the mass or mole balances for each component that is consumed (reactant) and produced (product) in the reactions.

It is especially favorable to set up balances on elements when multiple reactions occur in a system. Furthermore, previously discussed key elements are *key* to making the calculations of these types of systems easier.

Setting up mass balances of systems with chemical reactions will first be exemplified with the combustion of methane and then followed up with assignments for the reader to solve.

# 3.11 Practice Example: Excess Air Coefficient for a Natural Gas Boiler

**Problem statement:**

A company is planning to commission a natural gas boiler and has calculated the gas consumption to be 20 kmol per hour. To achieve complete combustion the $O_2$ content of the flue gas must be 3 mol% on dry-basis.

Calculate the excess air coefficient $\lambda$, assuming the natural gas is 100 % methane.

**Solution:**

This example will be solved following: 3.6 The Systematic Approach to Solving Mass Balances

1. The fuel consumption is given at 20 kmol/h. The theoretical air and the practical air needs to be calculated to get $\lambda$.

2. Draw a simple BFD with known and unknown variables

3. Add SI units to drawing

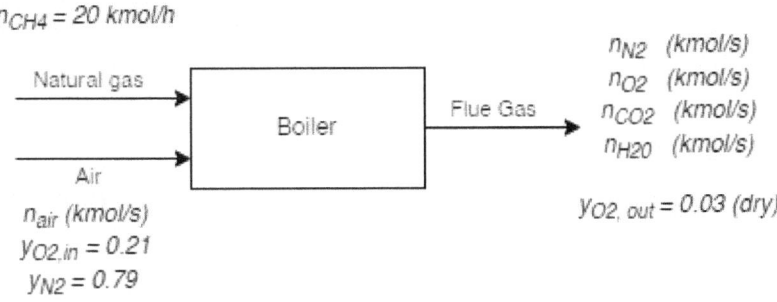

Figure 6: BFD - Combustion of methane in a boiler unit

4. System boundary is the boiler and the inlet and outlet streams.

5. Chemical reaction: $CH_4 + 2O_2 \rightarrow CO_2 + 2H_2O$

6. (+7) For element balances in this case there are 4 key elements: C, H, O, N.

With these 4 independent linear equations can be set up. There are 5 unknowns for this problem: $n_{air}, n_{N2}, n_{O2}, n_{CO2},$ and $n_{H2O}$. Conveniently, a condition-based equation can be set up knowing the oxygen composition of the flue gas has to be 3 mol%.

$$y_{O_2, out} = \frac{n_{O_2}}{n_{dry\ gas}} = \frac{n_{O_2}}{n_{O_2} + n_{CO_2} + n_{N_2}} = 0.03$$

Since the system is in steady state (Accumulation = 0) and elements cannot be consumed or produced (Produced = 0), the general mass balance for these can be reduced to:

$$IN = OUT$$

When setting up the element balances, account is taken for the amount of atoms in a given component for the element being balanced.

**Element balance for C:**

$$IN = OUT$$

$$n_{CH_4} \times 1\frac{kmol\ C}{kmol\ CH_4} = n_{CO_2} \times 1\frac{kmol\ C}{kmol\ CO_2}$$

$$n_{CO_2} = n_{CH_4} = 20\ kmol/h$$

**Element balance for H:**

$$IN = OUT$$

$$n_{CH_4} \times 4\frac{kmol\ H}{kmol\ CH_4} = n_{H_2O} \times 2\frac{kmol\ H}{kmol\ H_2O}$$

$$n_{H_2O} = 2 \times n_{CH_4} = 2 \times 20\ kmol/h = 40\ kmol/h$$

**Element balance for N:**

$$IN = OUT$$

$$n_{air} \times y_{N_2} \times 2\frac{kmol\ N}{kmol\ N_2} = n_{N_2} \times 2\frac{kmol\ N}{kmol\ N_2}$$

$$n_{N_2} = n_{air} \times y_{N_2} = n_{air} \times 0.79$$

**Element balance for O:**

$$IN = OUT$$

$$n_{air} \times y_{O_2} \times 2\frac{kmol\ O}{kmol\ O_2} = n_{CO_2} \times 2\frac{kmol\ O}{kmol\ CO_2} + n_{H_2O} \times 1\frac{kmol\ O}{kmol\ H_2O} + n_{O_2} \times 2\frac{kmol\ O}{kmol\ O_2}$$

$$n_{air} \times 0.21 \times 2 = 20\ kmol/h \times 2 + 40\ kmol/h + n_{O_2} \times 2$$

$$n_{O_2} = 0.21 \times n_{air} - 40\ kmol/h$$

Now the balance equations can be inserted into the condition-based equation:

$$y_{O_2, out} = \frac{n_{O_2}}{n_{dry\ gas}} = \frac{n_{O_2}}{n_{O_2} + n_{CO_2} + n_{N_2}} = 0.03$$

$$\frac{0.21 \times n_{air} - 40\ kmol/h}{0.21 \times n_{air} - 40\ kmol/h + 20\ kmol/h + n_{air} \times 0.79} = 0.03$$

Solving for n$_{air}$:

$$\frac{0.21\ n_{air} - 40\ kmol/h}{n_{air} - 20\ kmol/h} = 0.03$$

$$n_{air} = 218.9\ kmol/h$$

Then $n_{air}$ is inserted in the element balances to find the remaining unknowns needed to find the theoretical air:

$$n_{N_2} = n_{air} \times 0.79 = 172.9 \, kmol/h$$

$$n_{O_2} = n_{air} \times 0.21 - 40 \, kmol/h = 6 \, kmol/h$$

Now all that is left is to calculate the theoretical air, which is found by setting $n_{O2}$ equal to 0 in the element balance for oxygen (since theoretical air is the measure of total air where 100 % of the oxygen is used in the combustion.

$$n_{air,min} \times 0.21 \times 2 = 20 \, kmol/h \times 2 + 40 \, kmol/h + 0 \times 2$$

$$n_{air, min} = L_{min} \times 0.21 \times 2 = 80 \, kmol/h$$

$$L_{min} = 190.5 \, kmol/h$$

$$\lambda = n_{air}/L_{min} = \frac{218.9 \, kmol/h}{190.5 \, kmol/h} = 1.15$$

The calculations can then be controlled with a total mass balance:

$$IN = OUT$$

$$n_{CH_4}M_{CH_4} + n_{air}M_{air} = n_{CO_2}M_{CO_2} + n_{H_2O}M_{H_2O} + n_{O_2}M_{O_2} + n_{N_2}M_{N_2}$$

Using rounded molar masses the result is:

$$IN: 20 \times 16 + 218.9 \times 28.84 = 320 + 6348 = 6633 \, kg/h$$

$$OUT: 20 \times 44 + 40 \times 18 + 6.0 \times 32 + 172.9 \times 28 = 6633 \, kg/h$$

# 3.12 Assignment - Distillation of Methanol and Ethanol

## Problem statement:

A mixture of water, ethanol and methanol (stream 1) is distilled continuously in a distillation column, where a higher concentration of ethanol and methanol leaves via stream 2 and a stream mainly consisting of water exits at the bottom (stream 3).

Below is the BFD of the process:

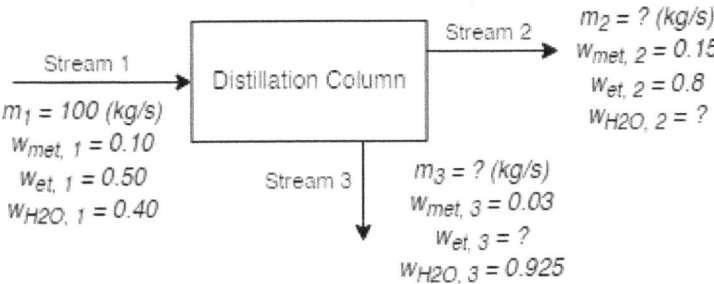

Figure 7: BFD - Distillation of methanol and ethanol

A. Calculate $w_{H2O}$ in stream 2 and $w_{et}$ in stream 3

B. Calculate the flow rates $m_2$ and $m_3$ by setting up the required number of equations in the form of mass balances.

---

## Solution:

## 6.3 Solutions to Assignments

---

# 3.13 Assignment - Production of Acetylene from Pyrolysis of Methane

**Problem statement:**

Acetylene can be produced from natural gas (pure methane in this example) via pyrolysis. The reaction is as follows:

$$2\,CH_4(g) \rightarrow C_2H_2(g) + 3\,H_2(g)\ (1)$$

The reaction is happening at high temperature (around 1500°C) and the required heat to drive the reaction is obtained from burning part of the natural gas. We assume that this combustion does not produce $CO_2$:

$$2CH_4(g) + 3\,O_2 \rightarrow 2\,CO\,(g) + 4\,H_2O(g)\ (2)$$

10 mol $CH_4$ is fed to the reactor per second. Via a second stream 5 mol concentrated $O_2$ (96 vol% $O_2$ and 4 vol% $N_2$) is added to the reactor. All of the oxygen is consumed in the combustion so there is no oxygen in the gas leaving the reactor.

The BFD is drawn below:

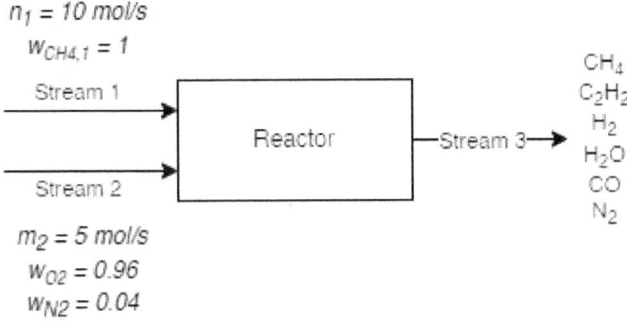

$n_1 = 10 \ mol/s$
$w_{CH4,1} = 1$

Stream 1

Reactor

Stream 3 →

CH$_4$
C$_2$H$_2$
H$_2$
H$_2$O
CO
N$_2$

Stream 2

$m_2 = 5 \ mol/s$
$w_{O2} = 0.96$
$w_{N2} = 0.04$

Figure 8: BFD - Production of Acetylene

A. Calculate how much $CH_4$(mol/s) that is used for the combustion (Reaction 2), and calculate the mole flows for $H_2O$ and CO leaving the reactor via stream 3.

B. Assume half of the added $CH_4$ is not consumed by the reactor and leaves with stream 3: Calculate how much $C_2H_2$ is being produced (mol/s).

---

**Solution:**

## 6.3 Solutions to Assignments

---

Now it's time to look at more complex systems and systems that are not in steady state.

# 4. Mass Balances for Complex & Non-Steady State Systems

Now that the fundamental concept of setting up mass balances is introduced, it's time to see how this applies to more complex and non-steady state systems which are more occurring in the real life of a process engineer.

After this chapter the reader should have achieved the following competencies:

- ❖ Describe what recirculation, bypass and purge entails.
- ❖ Account for the potential necessity of utilizing recirculation, bypass and purge in chemical processes.
- ❖ Apply correct system boundaries for use in setting up mass balances for steady state systems with and without chemical reactions.
- ❖ Setting up and solving mass balances for complex systems.
- ❖ Setting up and solving mass balances for non-steady state systems without chemical reactions.

# 4.1 Recirculation & Bypass

A chemical reaction does not always proceed fully from reactant to product and since costs of raw materials/reactants are usually a big chunk of the operational variable costs, it is often appropriate to separate unconverted reactant from the product stream and recirculate it back to the reactor vessel.

This is the reason that **recirculation** is an important concept to know as a process engineer.

**Bypass** is a similar concept, where the whole stream or a part of it is led around a unit to *bypass it.* Bypasses have a multitude of functions such as being able to bypass a filter before the product meets a certain standard, bypass a reactor to control the concentration or temperature of a product stream and many more.

**Purges** are used as recirculation comes with the side-effect of accumulating higher concentration of certain components in a stream over time. Purging is done either continuously or batch-wise to let a part of the recirculating stream exit the system.

Up until this point in the book, the type of systems that have been investigated have been simple systems. Simple systems are linear in nature.

Complex systems, which will be investigated in this chapter, involve the use of bypasses and recirculation, which immediately might seem daunting to apply mass balances on.

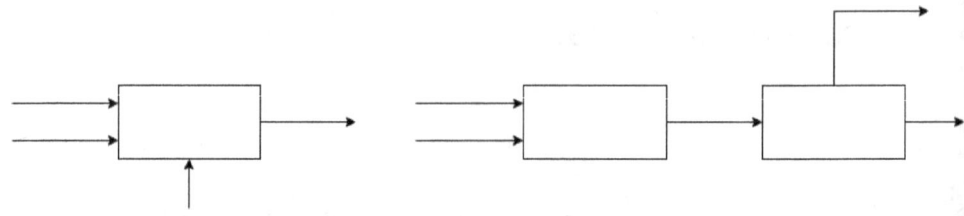

Figure 9: Examples of simple system BFDs

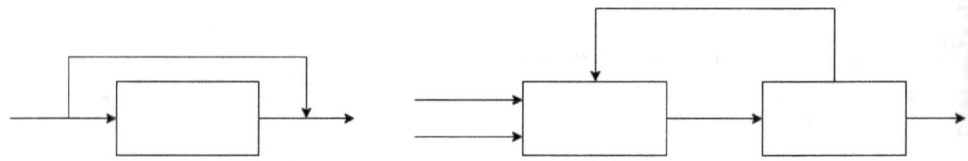

Figure 10: Examples of complex system BFDs

The key to solve the complexity of these types of systems lies in the system boundaries, and their ability to create simple sub-systems of a complex system.

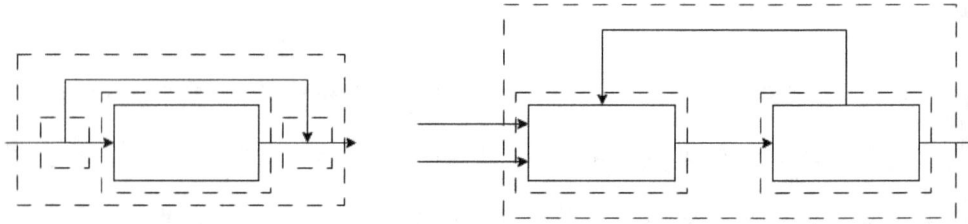

Figure 11: Complex systems divided into simple subsystems with system boundaries

Let's dive into an example straight away. Don't worry, it's not as difficult as it may seem on the surface.

# 4.2 Practical Example: Complex Cooling Loop System (Recirculation and Purge)

**Problem statement:**

Recirculating cooling loop systems are utilized to reduce costs on the cooling water requirement. The cooling water is heated in a heat exchanger and thereby cools a process stream. The cooling water is then re-cooled in a cooling tower, where it's running down on a grid of tree or plastic in counter-current with air, where part of the cooling water evaporates. This evaporation requires energy and the cooling water is thereby cooled.

During this recirculation and evaporation, naturally occurring salts are increasing in concentration in the cooling water. To avoid precipitation of the salts, part of the cooling water is purged into the sewer, at the same time as new water is added to the loop via the cooling tower.

Calculate the amount of water needed to be purged, when the maximum allowed concentration of $SiO_2$ is 150 mg/L.

The clean water added holds 5 mg $SiO_2$ /L. The flow rate through the heat exchanger is 158,000 m³/h. Assume a density of the cooling water at 1000 kg/m³ and that 1 % of the recirculating water evaporates in the cooling tower.

---

**Solution:**

*Note: For solving more complex systems like this one, really take the time to get the first stages right. Points 1-4 are key, since if mistakes are made in these steps, all other calculations will be wrong and time-consuming to correct. Creating the block diagram from text to visualize the system correctly is perhaps the most important step of this whole process.*

Let's again leverage the step-by-step guide from subchapter 3.6.

1. The flowrate of added fresh water and the purge stream to the sewer needs to be calculated. The concentration of $SiO_2$ of these two streams are known, and the amount of water that evaporates in the cooling tower can be calculated from the given information.

2. Draw a simple BFD with known and unknown variables

3. Add SI units to drawing

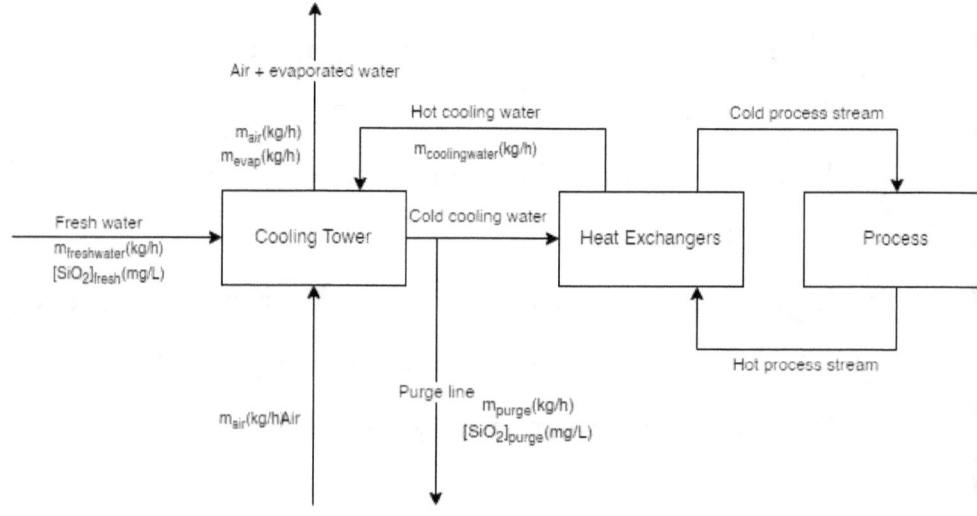

Figure 12: BFD - Complex Cooling loop without system boundaries

4.  Now that an overview of the process exists. With systems that include recirculations it's usually preferable to make the system boundary engulf the whole process.

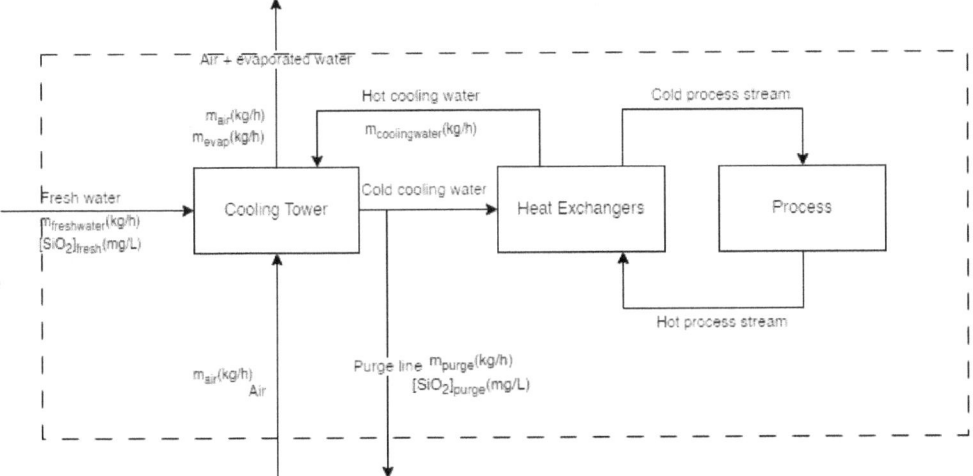

Figure 13: BFD - Complex Cooling loop with simplifying system boundaries

5.  There are no chemical reactions.

$$PRODUCED = 0$$

6.  +7. Setting up the mass balances.

First the flow rate of evaporative cooling water is calculated from the given information that 1 % of the water evaporates:

$$m_{vap} = m_{coolingwater} \times 0.01 = 158000 \, m^3/h \times \rho_{H_2O} \times 0.01$$

$$m_{vap} = 158000 \, m^3/h \times 1000 kg/m^3 \times 0.01 = 1,580,000 \, kg/h$$

For this problem there are 2 unknowns: mass flow rate of fresh water and mass flow rate of purged water. Therefore 2 balances are needed to solve this system.

The system is in steady state, which means the accumulation term is equal to 0. Since there are no chemical reactions, we can set up the total balance and component balance for $SiO_2$ with both Produced terms being equal to 0.

Since both terms of accumulation and produced are equal to zero, both balances can be written in the form of:

$$IN = OUT$$

**Total mass balance (kg/h):**

$$m_{fresh} + m_{air} = m_{purge} + m_{vap} + m_{air}$$

$m_{air}$ exists unaltered on both sides of the equation and drops out.

$$m_{fresh} = m_{purge} + 1,580,000 \ kg/h$$

**$SiO_2$-balance (kg/h):**

$$m_{fresh} \times [SiO_2]_{fresh} = m_{purge} \times [SiO_2]_{purge}$$

$$m_{fresh} \times 5 \ mg/L = m_{purge} \times 150 \ mg/L$$

$$m_{fresh} = m_{purge} \times \frac{150 \ mg/L}{5 \ mg/L} = m_{purge} \times 30$$

Using the reduced expression for $m_{fresh}$ and substitute it in the last expression from the $SiO_2$ balance:

$$m_{fresh} = m_{purge} + 1,580,000 \ kg/h = m_{purge} \times 30$$

$$m_{purge} = \frac{1,580,000 \ kg/h}{29} = 54,483 \ kg/h$$

$$m_{fresh} = m_{purge} \times 30 = 54483 \ kg/h \times 30 = 1,634,483 \ kg/h$$

$$m_{fresh} \approx 1630 \ m^3/h$$

In order to use 158,000 m$^3$/h cooling water, it is only needed to supply 1630 m$^3$/h fresh water to maintain an acceptable $SiO_2$ concentration. Most of the water added to the cooling loop is being evaporated and a relatively small amount is purged.

The loop recirculation and purge design thereby decreased the use of cooling water by 99 %, compared to a system with no recirculation and continuously added fresh water. This illustrates why recirculation loops are so abundantly found in the process industry.

---

This example hopefully shows the reader the power of structuring the approach and setting clever system boundaries. A problem that initially looked very complex was effectively reduced to two balances of the IN = OUT format.

# 4.3 Non-Steady State Systems

In chapter 2 of this book the general mass balance was introduced:

$$IN \ + \ PRODUCED \ = \ OUT \ + \ ACCUMULATED$$

In all calculations up to this point in the book the *accumulated* term has been equal to 0, as the systems considered were in steady state.

Non-steady state balances are used for processes where operational conditions or mass flow rates change over time. Practical examples of non-steady states are start-up and shut-down sequences of any steady state processes, significant changes from one set of operational conditions to another, or varying operational conditions.

In reality all complex industrial processes are never 100 % in steady state, as disturbance variables (operational conditions changing without intervention from an operator) will eventually shift the output of a continuous process if left uncontrolled. Process Control is an enormous field of engineering for this and many other reasons. Automatic stabilizing actions to stay as close to steady state as possible is usually very favorable.

So let's embrace that non-steady state is the reality of manufacturing and production (and the entire universe) and dive into a few examples.

The significant part of a non-steady state system is that the *accumulated* term of the general mass balance is **not equal** to 0.

The following general differential expression can be used to describe the changing nature of the total mass or mass of the component over time:

$$Accumulated\ total\ mass\ =\ \frac{d}{dt}(mass\ of\ the\ system)$$

or

$$Accumulated\ mass\ of\ component\ =\ \frac{d}{dt}(mass\ of\ component\ in\ the\ system)$$

To find numerical solutions for these types of systems we need to calculate or utilize given boundary conditions. This will be explained in more detail in the first example below. To ease into the world of non-steady state systems, let's look at an example of cleaning a vessel:

# 4.4 Practice Example: Cleaning a Vessel

**Problem statement:**

A vessel holds 500 L of a NaOH solution with a concentration of 1M (mol/L). The vessel needs to be cleaned. The cleaning is done by adding 2.0 L of water per second. The mixing is so effective that the concentration of NaOH can be assumed to be uniform throughout the vessel. 2.0 L is drained from the tank per second. The density of the contents of the vessel is assumed to be constant at 1.00 kg/L for the entire duration of the cleaning procedure. How much time does it take before the concentration of NaOH has dropped to 0.01M (0.01 mol/L). The total volume of fluid is assumed to be constant at 500 L, because of the equal inlet- and outlet flow rate of the vessel.

During the cleaning process, the concentration of NaOH is gradually reduced. The process is thereby not in steady state.

---

**Solution:**

1. Draw a diagram for the process including known and unknown variables with corresponding units

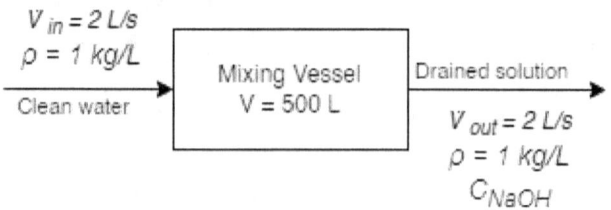

Figure 14: BFD - Cleaning mixing vessel (non-steady state system)

2. The drawing shows a system in non-steady state with one time-dependent variable: CNaOH.

3. Set up numeric boundary conditions for the time-dependent variable:

| t (seconds) | $C_{NaOH}$ (mol/L) | (mol/L) |
|---|---|---|
| 0 | 1 | $C_{NaOH}(0) = 1$ |
| t | $C_{NaOH}$ | $C_{NaOH}(t) = C_{NaOH}$ |
| $t_1$ | 0.01 | $C_{NaOH}(t_1) = 0.01$ |

4. Setting up a balance for NaOH (mol) per time in the tank:

No chemical reactions, PRODUCED = 0.

Non-steady state system ACCUMULATED ≠ 0.

$$IN = OUT + ACCUMULATED$$

$$0 = V_{out} \times C_{NaOH} + \frac{d}{dt}(V \times C_{NaOH})$$

The provided information is inserted:

$$0 = 2\,L/s \times C_{NaOH} + \frac{d}{dt}(500\,L \times C_{NaOH})$$

Rearranging to look like a standard differential equation:

$$\frac{d}{dt}(C_{NaOH}) = -\frac{2\,L/s}{500\,L}C_{NaOH}$$

The differential equation can be solved by separation of the variables[2]:

Dividing $C_{NaOH}$ and multiplying with dt on both sides of the equation:

$$\frac{dC_{NaOH}}{C_{NaOH}} = -\frac{2}{500\ s}dt$$

Now it's time for integration[3] with the given boundary conditions:

$$\int_{1}^{0.01} \frac{dC_{NaOH}}{C_{NaOH}} = -\int_{0}^{t_1} \frac{2}{500\ s}dt$$

$$[ln\ C_{NaOH}]_{1}^{0.01} = -\frac{2}{500\ s}[t]_{0}^{t_1}$$

$$ln\frac{0.01}{1} = -\frac{2}{500\ s}\ t_1$$

$$t_1 = -ln\frac{0.01}{1} \times \frac{500\ s}{2} = ln\ 100 \times 250\ s$$

$$t_1 = 1151\ s$$

It will take 1151 seconds before the concentration of NaOH has dropped to 0.01 mol/L.

---

Let's increase the complexity a bit by turning one of the constants into a time-dependent variable, namely the total volume of the tank.

---

[2] 6.2 Separation of Variables - Solving Differential Equations
[3] 6.1 Integral Rules

# 4.5 Practice Example: Cleaning a Vessel - Two Time-dependent Variables

## Problem statement:

The process is the same as for the previous example. The only changed variable is that 4 L of solution is drained per second instead of 2 L. This makes the volume of the tank also dependent on time. For this example let's find the concentration of NaOH once the vessel holds only 50 L.

## Solution:

1. Draw the diagram and set up the table for the boundary conditions:

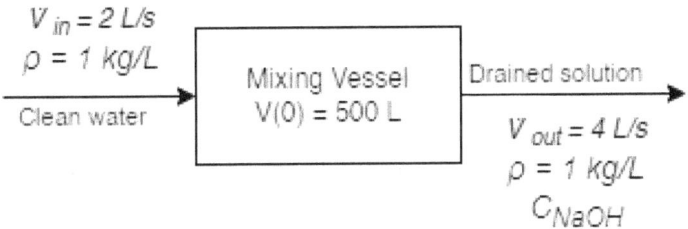

Figure 15: BFD - Cleaning mixing vessel (non-steady state system) (II)

| t (seconds) | $C_{NaOH}$ (mol/L) | V (L) | (mol/L) | (L) |
|---|---|---|---|---|
| 0 | 1 | 500 | $C_{NaOH}(0) = 1$ | $V(0) = 500$ |
| t | CNaOH | V | $C_{NaOH}(t) = C_{NaOH}$ | $V(t) = V$ |
| t1 | 0.01 | 50 | $C_{NaOH}(t_1) = C_2$ | $V(t2) = 50$ |

2. Set up the balances. Two balances are needed to solve the system of 2 time-dependent variables.

| **Balance:** | **IN** | **= OUT** | **+ ACC** |
|---|---|---|---|
| NaOH (mol/s) | 0 | $= V_{out} \times C_{NaOH}$ | $+ \frac{d}{dt}(V \times C_{NaOH})$ |
| Total mass (kg/s) | $V_{in} \times \rho$ | $= V_{out} \times \rho$ | $+ \frac{d}{dt}(V \times \rho)$ |

NaOH balance:

$$0 = 4 \times C_{NaOH} + \frac{d}{dt}(V \times C_{NaOH})$$

Total mass balance: (*densities cancel each other out*):

$$\frac{dV}{dt} = V_{in} - V_{out} = 4\,L/s - 2\,L/s = 2\,L/s$$

The resulting total mass balance of the system could have been written without calculations, but doing this is good practice for more complex problems.

Integration of the total mass balance expression by separation of variables[4] as done in the previous example:

$$\int_{500}^{V} dV = \int_{0}^{t} -2dt$$

---

[4] 6.2 Separation of Variables - Solving Differential Equations

$$V - 500 = -2t$$

Then we get the linear expression of emptying the tank:

$$V = 500 - 2t$$

This one could also have been written out just from simple logic, as the net-removal is very clearly 2 L per second and the start value is 500 L, however this is good practice to learn how to derive mathematically. Now that we have the linear expression for the tank volume we can insert this into our NaOH-balance:

$$0 = 4 \times C_{NaOH} + \frac{d}{dt}(V \times C_{NaOH})$$

$$0 = 4 \times C_{NaOH} + V\frac{dC_{NaOH}}{dt} + C_{NaOH}\frac{dV}{dt}$$

$$0 = 4 \times C_{NaOH} + (500 - 2t)\frac{dC_{NaOH}}{dt} + C_{NaOH}\frac{d(500-2t)}{dt}$$

$$0 = 4 \times C_{NaOH} + V\frac{dC_{NaOH}}{dt} - 2C_{NaOH}$$

$$-2C_{NaOH} = (500 - 2t)\frac{dC_{NaOH}}{dt}$$

$$\frac{dt}{500-2t} = -\frac{1}{2}\frac{dC_{NaOH}}{C_{NaOH}}$$

After this separation of variables, the expression can be integrated using the boundary conditions:

$$\int_{0}^{t_2} \frac{dt}{500-2t} = -\frac{1}{2}\int_{1}^{C_2} \frac{dC_{NaOH}}{C_{NaOH}}$$

Since $V(t_2)$ is 50 L, we can calculate $t_2$ from our linear equation:

$$V(t_2) \; = \; 50 \; = \; 500 - 2t_2$$

$$t_2 \; = \; \frac{500-50}{2} \; = \; 225 \; seconds$$

The left hand side of the integral[5] can now be calculated:

$$\int\limits_{0}^{225} \frac{dt}{500-2t} = -\frac{1}{2} \int\limits_{0}^{225} \frac{d(500-2t)}{500-2t} \; = \; -\frac{1}{2} [ln(500 - 2t)]_0^{225}$$

$$-\frac{1}{2}(ln(500 - 450) - ln(500)) \; = \; -\frac{1}{2} ln(\tfrac{50}{500}) \; = \; -\frac{1}{2} ln(0.1)$$

Inserting the left hand side into the equation with both integrals:

$$\int\limits_{0}^{t_2} \frac{dt}{500-2t} = -\frac{1}{2} \int\limits_{1}^{C_2} \frac{dC_{NaOH}}{C_{NaOH}}$$

$$-\frac{1}{2} ln(0.1) = -\frac{1}{2} \int\limits_{1}^{C_2} \frac{dC_{NaOH}}{C_{NaOH}} = -\frac{1}{2} [ln \, C_{NaOH}]_1^{C_2}$$

$$ln(0.1) \; = \; [ln \, C_{NaOH}]_1^{C_2} \; = \; ln \, C_2 - ln \, 1 \; = \; ln \, C_2$$

$$C_2 \; = \; 0.1$$

The concentration of NaOH when there is 50 L left in the vessel is 0.1M (0.1 mol/L).

---

[5] 6.1 Integral Rules

# 4.6 Assignment: Production of Methanol

**Problem statement:**

In a chemical reactor at high pressure and temperature a mixture of $H_2$ and CO are converted to $CH_3OH$ following the stoichiometric reaction equation:

$$CO + 2H_2 \rightarrow CH_3OH$$

In the reactor only 50 % of the CO is converted to $CH_3OH$. The rest passes through the reactor. The produced methanol is separated in a chiller-separation unit, where methanol is extracted in liquid form. The unconverted gas is recirculated from the separation unit and mixed back into the feed stream that enters the reactor.

The fresh feed stream (1 - before the recirculation tie-in), has a flow rate of 300 kmol/h with stoichiometric ratio of $H_2$ and CO. The BFD is drawn below:

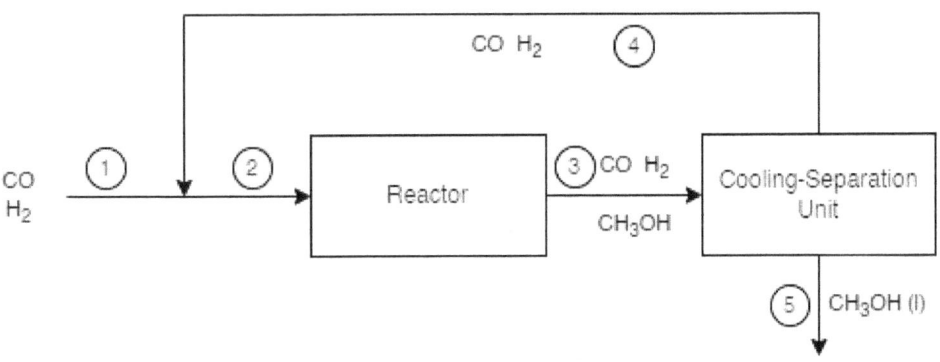

Figure 16: BFD - Production of Methanol with recirculation of reactants

**Fill in the remaining flow rates in the table below:** (Units in kmol/h)

| Component | Stream 1 | Stream 2 | Stream 3 | Stream 4 | Stream 5 |
|-----------|----------|----------|----------|----------|----------|
| $H_2$ | 200 | | | | 0 |
| CO | 100 | | | | 0 |
| $CH_3OH$ | 0 | 0 | | 0 | |
| SUM TOTAL | 300 | | | | |

---

## Solution:

## 6.3 Solutions to Assignments

---

# 4.7 Assignment: Production of Vegetable Oils

## Problem statement:

Dried beans are assumed to consist of 10.0 w% vegetable oil, and 90 % dry matter (that is not soluble in n-hexane($C_6H_{14}$)).

The beans are pre-treated and are added to the extraction tank along with n-hexane in the ratio of 2 kg beans per 3 kg n-hexane (2:3). The n-hexane that enters the extraction tank is a mix of a fresh stream and recirculated stream of n-hexane.

The slurry from the extraction tank is led to a filtration unit. The filter cake exiting this unit is 75 w% dry matter and 25 w% mixture of n-hexane and vegetable oil. The composition of this mixture of n-hexane and oil is the same in the slurry, the filter cake and the filtrate, that is lead to the evaporator.

In the evaporator the mixture is split in n-hexane (g) and vegetable oil (l).

The gaseous n-hexane is then led to a condensation unit, wherefrom the liquid n-hexane is then mixed with the fresh n-hexane before it enters the extraction tank.

The BFD is drawn below:

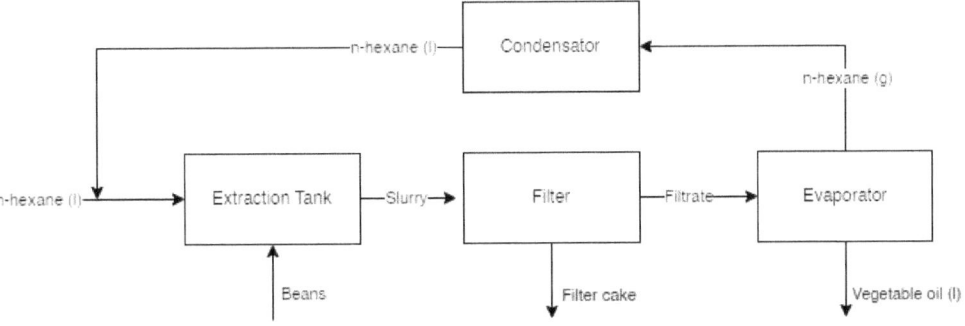

Figure 17: BFD - Production of vegetable oil with recirculation of n-hexane

A. Show that the ratio of the mass fractions of n-hexane and vegetable oil is 15:1.
B. Calculate the yield of vegetable oil as kg oil / kg added beans.
C. Calculate the needed amount of fresh n-hexane per kg added beans.
D. Calculate the ratio of recirculated- and fresh n-hexane.

---

**Solution:**

### 6.3 Solutions to Assignments

---

# 5. Conclusion

*Well done!*

Throughout this book, you have explored some of the most important and foundational process engineering concepts such as process diagrams, system types and mass balances of simple & complex steady state and non-steady state systems with and without chemical reactions.

By following the practice examples and doing the assignments, you have gained a **deeper understanding** of the concepts and have learned how to apply them to **solve real-world engineering problems**.

My hope is that the knowledge gained through this book will empower you to approach complex engineering problems with confidence and provide you with a solid foundation for further study and work in the field of process engineering.

*Well wishes for the future,*
**Michael Kay Hoffmann**

# 6. Appendices

## 6.1 Integral Rules

| Common Functions | Function | Integral |
|---|---|---|
| Constant | $\int a\ dx$ | $ax + C$ |
| Variable | $\int x\ dx$ | $x^2/2 + C$ |
| Square | $\int x^2\ dx$ | $x^3/3 + C$ |
| Reciprocal | $\int (1/x)\ dx$ | $\ln|x| + C$ |
| Exponential | $\int e^x\ dx$ | $e^x + C$ |
| | $\int a^x\ dx$ | $a^x/\ln(a) + C$ |
| | $\int \ln(x)\ dx$ | $x\ \ln(x) - x + C$ |
| **Trigonometry** | $\int \cos(x)\ dx$ | $\sin(x) + C$ |
| | $\int \sin(x)\ dx$ | $-\cos(x) + C$ |
| | $\int \sec^2(x)\ dx$ | $\tan(x) + C$ |

| Rules | Function | Integral |
|---|---|---|
| Multiplication by constant | $\int cf(x)\ dx$ | $c\int f(x)\ dx$ |
| Power Rule ($n \neq -1$) | $\int x^n\ dx$ | $\dfrac{x^{n+1}}{n+1} + C$ |
| Sum Rule | $\int (f + g)\ dx$ | $\int f\ dx + \int g\ dx$ |
| Difference Rule | $\int (f - g)\ dx$ | $\int f\ dx - \int g\ dx$ |

# 6.2 Separation of Variables - Solving Differential Equations

Example walkthrough:

Given an equation of:
$$\frac{dy}{dx} = \frac{2xy}{1+x^2}$$

and the boundary conditions: $x_0$, $x_1$, and $y_0$, $y_1$

Step 1: Isolate all the y and dy terms to one side of the equation and all the x and dx terms to the other:

$$\frac{dy}{dx} = \frac{2xy}{1+x^2}$$

$$\frac{dy}{y} = \frac{2x}{1+x^2}dx$$

$$\frac{1}{y}dy = \frac{2x}{1+x^2}dx$$

Step 2: Integrate one side with respect to y and the other side with respect to x using the boundary conditions given or calculated from the problem statement:

$$\int_{y_0}^{y_1}\frac{1}{y}dy = \int_{x_0}^{x_1}\frac{2x}{1+x^2}dx$$

$$[ln\,y]_{y_0}^{y_1} = [ln(x^2 + 1)]_{x_0}^{x_1}$$

Step 3: Simplify the expression:

$$ln\,y_1 - ln\,y_0 = ln(x_1^2 + 1) - ln(x_0^2 + 1)$$

$$y_1 - y_0 = (x_1^2 + 1) - (x_0^2 + 1)$$

# 6.3 Solutions to Assignments

## 3.11 Assignment - Distillation of Methanol and Ethanol
a.    $w_{H2O,2} = 0.05$; $w_{et.3} = 0.05$
b.    $m_2 = 60$ kg/h; $m_3 = 40$ kg/h

---

## 3.12 Assignment - Production of Acetylene from Pyrolysis of Methane
a.    3.2 mol $CH_4$ is combusted per second, $n_{CO} = 3.2$ mol/s; $n_{H2O} = 6.4$ mol/s
b.    0.9 mol $C_2H_2$ is produced per second.

---

## 4.6 Assignment: Production of Methanol

(Units in kmol/h)

| Component | Stream 1 | Stream 2 | Stream 3 | Stream 4 | Stream 5 |
|---|---|---|---|---|---|
| H2 | 200 | 400 | 200 | 200 | 0 |
| CO | 100 | 200 | 100 | 100 | 0 |
| $CH_3OH$ | 0 | 0 | 100 | 0 | 100 |
| SUM TOTAL | 300 | 600 | 400 | 300 | 100 |

---

## 4.7 Assignment: Production of Vegetable Oils
a.    15 : 1
b.    0.081 kg oil / kg beans
c.    0.28 kg kg fresh n-hexane / kg beans
d.    Recirculation ratio for n-hexane = 4.3